Fokker V 5/Dr. 1
by Wolfgang Schuster and Achim Sven Engels

Why the three wings!?

The idea of fitting an aircraft with three pairs of wings was not a product of aircraft development during the First World War, but instead hearkened back to the year 1843. At that time Sir William Samuel Henson (1805 - 1888), together with his partner John Stringfellow, published in an edition of Mechanic's Magazine details of a heavier than air vehicle which was to have been driven by a powerful steam engine and have incorporated a wingspan of over 45 meters. This article was read by Sir George Cayley (1773 -1857). In a letter to the editor which appeared in the same magazine, Cayley wrote that it would be more sensible to shorten the wingspan of the apparatus and instead fit it with three wing pairs, one on top of the other. This was probably the first time in the history of aviation that a lifting body was mentioned which had three wings.

Sir Cayley's recommendation was picked up by John Stringfellow, who subsequently began experimenting with models of triplanes. These experiments were not successful, however, and he eventu-

ally gave up on the idea. The first successful flight of a motor-driven triplane occured in 1908, when Ambroise Goupy in his Goupy 1 flew for distances of up to 150 meters. Such performance from his triplane attracted considerable attention from the aviation world at the time. By the end of 1908 Farman had built a triplane, which on 6 January 1909 was able to fly for a distance of 8 kilometers near Bouy.

At about the same time, Hans Grade succeeded in flying a distance of 400 meters in his small triplane in Germany.

In Great Britain it was Alliot Verdon-Roe who conducted trials with his triplanes and used the results to make constant improvements to his machines. In 1910 he founded the A.V. Roe Company, which is known today for its aircraft designs under the name of AVRO.

The Triplane Concept in World War I

The triplane concept found new fame in the middle of the First World War when the British firm Sopwith on Thames put its newest project, the LC1TTr (known as the Sopwith

Triplane), into action against the German Fliegertruppe. In the experienced hands of the Royal Naval Flying Corps (RNFC) this airplane was such a successful weapon that it can truthfully be said it posed a significant danger to every single German combat aircraft at the time. The Inspektion der Fliegertruppe(Idflieg, or Inspector's Office of the Flying Corps) was of the opinion that the British, with the triplane concept, had come up with a magical formula for building single-seat fighters. On 27 July 1917 Hauptmann Mühlig-Hoffmann was assigned by senior aviation personnel to prepare a paper, which was subsequently disseminated to all German aircraft manufacturers. The contents of the letter let it be known that a captured Sopwith Triplane was in Adlershof, the German aircraft test center, and would be made available for examination by company engineers. At the same time the aviation companies were informed that the Idflieg would in principle support all efforts in the development of practical triplane fighters as soon as such projects promised success.

Tubular steel frame of triplane, basic structure. Factory photo taken at the Fokker Flugzeugwerke GmbH.

The result of Idflieg's proposal was that virtually all German fighter manufacturers tripped over themselves to "develop" triplane aircraft. It is highly doubtful whether the majority of these companies were in fact serious about the developmental work involved. In looking at some of the products, one wonders whether these would have ever been truly considered for production, or whether the manufacturers were actually attempting to garner a developmental contract as quickly as possible, simply producing a design which met the basic requirements and was fitted with three wings.

At first, biplanes which had more or less proven to be reliable had their airframes modified to take a third set of wings. For the most part, their resulting performance was naturally found to be questionable. Several companies left it at that, while others continued to devote serious work to the program in an attempt to improve their products. Altogether, 34 various triplanes were eventually built by German manufacturers through the end of 1918.

Amazingly, only two aircraft manufacturers succeeded in producing a practical prototype: the Pfalz Flugzeugwerke GmbH in Speyer am Rhein and the Fokker Flugzeugwerke GmbH in Schwerin/Mecklenburg. Both of these aircraft were, given the airframe structure, entirely new designs and were not based on any existing aircraft.

The Pfalz Dr.I

In October 1917 the Pfalz Dr.I completed its testing phase at Adlershof. During the prototype's acceptance tests it was discovered that the Pfalz, with its 160 hp Siemens-Halske rotary engine, had a significantly better performance - particularly with regard to its rate of climb - than that of the Fokker triplane, and those within the Idflieg hoped for deliveries of 300 Pfalz Dr.Is as soon as possible. This, however, was not to be the case. The Pfalz Flugzeugwerke was awarded an initial contract for ten triplanes in October of 1917, to be used for combat trials at the front. The front-line pilots quickly gained experience with the new type, and just as quickly learned that the aircraft's new engine, the SH III, had not yet reached operational maturity and was the root of many performance problems. These problems had a negative impact on the flight handling characteristics of the triplane. In April of 1918 nine of the ten Pfalz Dr.Is which had been ordered were in action on the front lines. Since their flight characteristics were not particularly good due to the choice of engine, the majority of these airplanes were not put into action in the most important front-line sectors. Many of them saw service with Jagdstaffel 73.

Pfalz internally worked on improving the Dr.I as a means of addressing the prob-lems. This resulted in the Dr.II and Dr.IIa, neither of which was demonstrated for Idflieg.

The Dr.II was a smaller version of the Dr.I and was fitted with the reliable Oberursel Ur II nine-cylinder rotary engine with an output of 110 hp. Compared with the Dr.I, the Dr.II had its wingspan shortened by 1.35 m, the fuselage lengthened by 45 cm and its overall weight reduced by 110 kg. Rate of climb for the Dr.II was significantly worse than that for the Dr.I. The Dr.I took just 14.5 minutes to climb to 6,000 meters, whereas the Dr.II required 29 minutes to reach the same altitude.

Since the overall performance picture of the Dr.II was much worse (i.e. there was no improvement whatsoever over the Dr.I), an attempt was made to improve performance by mating the Siemens-Halske SH I nine-cylinder rotary engine - lighter by 28 kg - to the airframe. However, this approach was a failure as well; virtually nothing changed with regard to the poor climb rate of the airframe.

Neither the Dr.II nor the Dr.IIa went into production. The fate of the Pfalz triplane was sealed, and Idflieg's decision was made in favor of the triplane produced by the Fokker Flugzeugwerke, which by that time was already appearing in small numbers with front-line units.

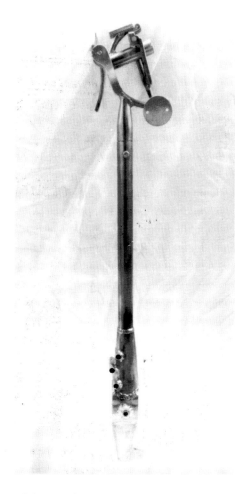

Side view of the Fokker Dr.I's control column prior to finishing. It was made of welded tubular steel.

Fokker's Path to a Triplane Design

Anthony Herman Gerard Fokker (1890 - 1939) occasionally visited the front lines himself in order to talk with Jagdstaffel pilots about the advantages and disadvantages of both his and competing designs. He was the only aircraft manufacturer to do this personally, and because of this was famous among the pilots for always having an open mind with regard to their wants and needs. As was the case around the end of April 1917 when he visited Jasta 11 and its squadron leader, Manfred Freiherr von Richthofen (1893 -1918). Von Richthofen had recently (on 20 April 1917) had a nerve-wracking experience with a Sopwith Triplane during which the enemy pilot conducted a brutal cat-and-mouse game with him, narrowly surviving the encounter in his Albatros D. III. As a result, he subsequently impressed upon Fokker the urgency in developing a high-performance airplane which could hold its own with the new British combat aircraft. On this occasion Fokker was also probably shown one of the first Sopwith aircraft to have been captured intact prior to its being shipped off to Adlershof. At the same time Richthofen invited Fokker to witness first-hand one of these triplanes in a dogfight from the vantage point of an observation post on the front lines.

It has never been fully explained how Fokker reacted to Richthofen's request at the time. What is known, however, is that on 3 June 1917 he assigned his developmental bureau a contract for the design of a biplane which was expected to enter service with the Austro-Hungarian Fliegerkorps. This biplane, designated the Fokker D.VI and given Werknummer (Wk.Nr. or factory aircraft serial number) 1661, was converted into a triplane during the developmental phase on Fokker's orders. What exactly caused him to do this has also never been adequately explained. We believe that the reason for this sudden change of mind is found in the fact that Fokker knew that both the Siemens-Schuckert Werke and the Pfalz Flugzeugwerke were about to sign agreements with the Idflieg for the construction of three triplane prototypes each, to be financed by the Idflieg.

By June of 1917 test flights were underway using the triplane Fokker D.VI, during which period the technical data was assimilated and logged. This was confirmed by a British secret service report dated 2 July.

The aircraft's center wing and lower wing both had the same wingspan, a wing chord of 980 mm, but there were no aerodynamic counterbalances for the ailerons or horizontal stabilizers and the wings were lacking any type of support struts between them.

However, the British secret service report also contained some information which was erroneous. For example, the wingspan of the center and lower wings was given as different, the engine was supposedly either a 160 hp Daimler six-cylinder inline type or one produced by the Motorenfabrik Oberursel Aktiengesellschaft. Armament was mentioned as being two LMG 08/5 machine guns. This description could in no way have pertained to Wk.Nr. 1661, since this airframe had neither armament nor the 160 hp Daimler engine, and the wingspan of both the center and lower wings was the same, as mentioned earlier.

It is conceivable that, shortly before the secret report was compiled, the original prototype of the Fokker D.VI (Wk.Nr. 1661) was modified in such a way that the center and lower wings had different wingspan dimensions, so that the wingspan increased in equal proportions from the bottom wing to the top wing.

In fact, the wingspan of the center wing was indeed lengthened. At the same time, the ailerons and horizontal stabilizers were fitted with their distinctive counterbalances to enable the pilot to more easily maintain control. Additionally, the wings were joined together using airframe struts which at the same time prevented them from fluttering. These modifications also included fitting the airframe with machine guns.

After the D.VI (Wk.Nr. 1661) had been test flown by Fokker in its original form in Schwerin and the Idflieg, probably through Bauaufsicht 13 (construction supervision department 13) within the Fokker Werke, had learned of the aircraft's performance, for the first time it began to show increased interest in the design.

This is probably also a shot of 101/17. This is a rare in-flight photograph, viewing the triplane almost directly from the front.

Anthony Herman Gerard Fokker in the cockpit of a triplane, conversing with General-Major von Lossberg. The photo was taken during the time when the type was shown to and flown before leaders of the IV Army. It was taken at Jasta 11's airfield at Marckebeeke and shows one of the two pre-production triplanes, Fok. F.I 102/17.

Leutnant Werner Voss (1897 - 1917), a young flying ace and friend of Fokker, also had the opportunity to fly the D.VI (Wk.Nr. 1661) in Schwerin. It is possible that the modifications were undertaken based on his feedback. Although he found the flight characteristics outstanding, particularly with regard to the aircraft's climb rate and maneuverability, he was concerned that the flutter effect of the unbraced wings would lead to anxiety among pilots on the front.

The modified D.VI (Wk.Nr. 1661) was sent to Matyasföld in Hungary at the end of August 1917, where it was delivered to the Allgemeine Ungarische Maschinenfabrik

AG so that it could be demonstrated for Flars (the Austro-Hungarian equivalent of Germany's Idflieg). Unexpectedly for Fokker, however, Austria placed no orders for the type.

On 5 July 1917 Fokker assigned his developmental bureau with the development of a second triplane, also designated as D.VI. This machine was given to Werknummer 1697 (later 101/17). It was probably built due to the fact that at this point in time Fokker's financial situation was relatively poor, and there was urgent need for an immediate official contract for large-scale series production. He was therefore

trying to present to the Idflieg a feasible triplane as soon as possible, preferably before the competition could beat him to it.

In accordance with his company policy up to that point of building new prototypes in two variants, one with a lighter, lower-powered engine and the other with a heavier, high-power engine, on 7 July 1917 he assigned his developmental bureau to develop an inline version of the D.VI. This variant was designated the D.VII and given Werknummer 1698. With an all-up weight of 880 kg it proved to be too heavy and, after minor modifications and short test flights, was eventually dropped due to its sloppy handling characteristics.

Once the D.VI triplane (Wk.Nr. 1697) had been approved by the Idflieg, Fokker received the news that he was scheduled for formal prototype evaluation at the earliest date possible. Since the Idflieg's usual policy was to acquire a total of three prototypes for formal evaluation - one for stress testing and the other two to be sent to the front for combat testing - on 11 July 1917 Fokker began building two additional triplanes in his factory (Wk.Nr. 1729 and 1730). For the first time these two triplanes began appearing in official Fokker company records with the prototype designation of "V" 4. The old prototype designation of "M" hadn't been used since the end of 1916 with the M. 22 (military designation D. V). At first, no new designation for prototypes was introduced. New types initially were given the designation of "D". This, however, was normally the designation of aircraft which had already been accepted for production by the military. Following the M. 22, developmental work on this was undertaken by Fokker. From the beginning he worked on achieving a cantilever wing construction for his prototypes. The D.VI triplane (Wk.Nr. 1661) was the fourth such Fokker prototype to have this design feature.

The wreckage of 102/17. This crash cost the life of Oberleutnant Kurt Wolff on 15 September 1917.

A very nice shot of Leutnant Heinrich Gontermann's 115/17. This photo clearly shows the large aerodynamic counterbalance of the right aileron.

Once it became apparent that Moser's concept would lead to a production contract, an internal decision was reached within Fokker to introduce the letter "V" as a designator for the new prototypes. The three previous Platz designs were also retroactively designated in the same manner (V. 1 through V. 3).

As early as 14 July 1917 the Idflieg placed an order for 20 Fokker triplanes which was to not only include those aircraft already under construction, but also the three previously built machines. This production batch was to collectively be assigned the aircraft classification of "F", a temporary class only just introduced.

After its construction the D.VI triplane (Wk.Nr. 1697) was shipped to Adlershof in order to carry out basic structural evaluation to be followed by stress testing. From this point on it carried the designation of Fokker F.I 101/17. The evaluation program began on 7 August 1917 and lasted until the 9th of August. Stress testing, which as a rule resulted in the aircraft being subjected to loads until it eventually broke, took place on 11 August 1917. In the official documents compiled by the Flugzeugmeisterei during the type evaluation, the D.VI (Wk.Nr. 1697) 101/17 was already being designated as the Fokker Dr.I. This is somewhat remarkable insofar that the "Dr" aircraft classification only officially went into effect on 19 August 1917 on the orders of the Flugzeugmeisterei (Abt.

AB Nr. 357701). From now on, all Fokker triplanes with the exception of 102/17(Wk.Nr. 1729, built on 16 August) and 103/17 (Wk.Nr. 1730, also completed on 16 August) carried the designation of Fokker Dr.I on their fuselage sides.

The two Fokker F.I triplanes, 102/17 and 103/17, were delivered to the front following their construction in Schwerin and subsequent acceptance by the Zentrale Abnahmekommission (ZAK, or Central Acceptance Commission). Although they were accepted as Dr.I types, they retained their

Fokker F.I designation on the fuselage. The Fokker Werke had used this designation prior to the introduction of the new "Dr" aircraft classification and didn't bother to change it afterwards. Fokker F.I 102/17 was delivered to the commander of Jagdgeschwader I, Manfred Freiherr von Richthofen, and assigned to Jasta 11 based in Marckebeeke in occupied Belgium. The second triplane, Fokker F.I 103/17, was presented to the squadron leader of Jasta 10, Leutnant Werner Voss, also based in Marckebeeke.

Taken on the same day and from virtually the same angle, this is also a photo of Leutnant Heinrich Gontermann's 115/17.

Combat testing began with the arrival of both prototypes at Jagdgeschwader I. However, bad weather delayed the first flights of the two triplanes until the 28th of August, 1917.

Also, at the end of August, the airfield at Marckebeeke was host to a demonstration of the new fighters, put on for the leaders of IV Army, to which Jagdgeschwader I was subordinated. The high-ranking army officers were visibly impressed with the good flight performance. On 15 September 1917 Leutnant Kurt Wolff, leader of Jasta 11, fell in combat while flying Fokker F.I 102/17. Just a few days later, on 23 September 1917, the same fate befell Leutenant Werner Voss in 103/17 when he became embroiled in a dogfight with seven British SE 5a aircraft from the RAF. He fell to his death near Frenzenberg behind British lines. His triplane became the subject of a British investigation report.

As already mentioned, the first production batch of triplanes included prototypes 101/17, 102/17 and 103/17. Subsequent machines from the first order were modified somewhat from the original design. These modifications included fitting the underside of the lower wings with curved skids and straightening the leading edge of the horizontal stabilizer. Probably the biggest change, however, was increasing the wing chord by 20 mm, from 980 mm to 1000 mm.

The wreckage of Gontermann's triplane after the accident in La Neuville. This photo also shows the fatal design of the aileron's aerodynamic balance to good effect.

The three prototypes had been built from standard 1:10 scale drawings, but once the ZAK had accepted the design proper factory plans were drawn up. A blueprint from 15 August 1917 confirms that the wing chord had already been changed to 1,000 mm by this date. Work began on 104/17 (Wk.Nr. 1772) on 13 September 1917 with this new measurement, while the prototypes retained the 980 mm depth (a fact confirmed by the documents prepared by the Flugzeugmeisterei during the aircraft's type testing phase).

Another shot of the remains of 115/17. Gontermann died the following day as a result of his severe injuries. This wreckage, together with Pastor's 121/17, was examined in detail in an attempt to determine the cause of the accidents.

These modifications led to the internal Fokker designations of V.4 and V.5. For the time being, this was an end to the developmental program at Fokker, although there were test prototypes of triplanes built later for evaluation purposes.

The Fokker Dr.I at the Front

Actual series production of the Fokker triplane began on 13 September 1917 with the construction of 104/17 (Wk.Nr. 1771). At the end of September Idflieg ordered the second delivery batch of triplanes (121/17 - 220/17). The first triplane to be delivered to the front was 115/17. With this aircraft, the new fighter was introduced into combat in service with VII Army. It reached Jasta 15 on 11 October 1917 and was assigned to that unit's squadron leader, Leutnant Heinrich Gontermann. On 10 October 1917 the next eleven machines (104/17, 106/17, 107/17, 109/17, 110/17, 111/17, 112/17, 113/17, 114/17, 116/17 and 118/17) rolled out of the Fokker assembly hangars, destined for Jasta 11. On 13 October 1917 these eleven were followed by an additional six machines (119/17, 121/17, 122/17, 123/17, 125/17 and 132/17), which were also assigned to Jasta 11. All seventeen of these aircraft reached their intended destination together on 20 October 1917.

As there were already several aircraft in active service on the front lines, on 25 October 1917 the Idflieg randomly selected a production aircraft and ordered that it be brought to Adlershof to determine if production machines actually matched the performance figures obtained during the prototype evaluation (7 - 11 August 1917) and which had formed the basis for the triplane's acceptance contracts. The choice fell upon 141/17.

Due to the poor weather conditions along the front in VII Army's sector, Leutnant Gontermann was prevented from taking his 115/17 aloft until 26 October 1917. On 28 October 1917 parts of the upper wing fell off while the aircraft was flying at 700 meters over the airfield of La Neuville, leading to the crash and total write-off of 115/17. Leutnant Gontermann suffered serious injuries and died as a result the following day. On 31 October 1917 Leutnant Günther Pastor and his 121/17 suffered the same fate as Leutnant Gontermann and his 115/17 three days prior. He crashed one kilometer north of Moorseele and was killed. After these tragic accidents, which were obiously not due to any enemy fire, all Fokker Dr. Is were grounded. Accordingly, pilots along the front had to return to flying their antiquated Albatros D.Vs and Pfalz D.IIIa types in order to maintain the flying tempo. The ZAK brought the so-called Sturzkommission (Crash Commission) on board and assigned it the responsibility of finding the cause of the accidents.

Leutnant Günter Pastor suffered the same fate as Leutnant Gontermann, when on 31 October 1917 under virtually the same circumstances he crashed and was killed in 121/17.

Remains of 121/17 seen from a different angle. After Pastor's crash the triplane was grounded and the Sturzkommission was tasked with finding the cause of the accidents.

Investigation of the Sturzkommission and its Results

On 2 November 1917 the Sturzkommission began its investigation into both accidents by examining the wrecks of both 115/17 and 121/17 as well as taking statements provided by eyewitnesses. The Sturzkommission's initial findings were made available on 4 November 1917. In brief, these showed that the aerodynamic counterbalance of the aileron was too large in size and subsequently became so contorted during certain types of maneuvers that the resulting air flow tore the aileron's attachment spar from its mounts.

But the deformation of the large aerodynamic balance alone was not the sole cause of the crash. Extremely poor workmanship in the Fokker Werke also played a role in the loss of the two fighter pilots and their machines. In order to determine with certainty whether this was the case, the aircraft with the greatest number of flying hours (according to Richthofen) had the skin removed from its upper wing, the spars and ribbing of which were examined in detail. This investigation showed that the internal structure of the wing, which made use of glue to hold it together, had become so damaged by moisture that virtually all the glued surfaces had separated. Examination of two additional triplanes, which at this time had the fabric from their lower wings removed for inspection as well, solidified this perception. Many of the rib gussets had already completely separated from the web ribs or spars.

Another point of criticism was that the fabric covering was pinned to the rib gussets. Several of the pins had not only missed the web rib, but also the rib gussets as well.

As a result of the investigation findings up to that point, the Sturzkommission determined that the Fokker triplane was unsuitable for front-line service. In doing so, it officially confirmed the formerly temporary grounding of the type. At the same time, the Fokker company was to undertake a reworking of the wing design, strengthening it where needed. This was to be accomplished at Fokker's own expense.

During the course of the investigation, the aileron of the Fokker Dr.I in Adlershof (141/17) was subjected to load testing on 5 November 1917. Amazingly, the results of this testing showed that the aileron withstood all stress loads with bravado. As a result of the fact that the aileron's design was sound in and of itself, the Idflieg sent a memorandum to Fokker on 6 November 1917 in which the required modifications to the wing design were listed. Only after these were incorporated would the triplane once again be cleared for front-line operations. The letter contained the following twelve points:

127/17 was reputedly one of the triplanes flown by Manfred Freiherr von Richthofen. The aircraft belonged to Jasta 11. Notice that the white background of the rudder has been overpainted in olive green, leaving just a 5 cm white border around the cross marking.

1. Reinforcing the wingtips by adding a supplemental rib to form a box rib design.
2. A third box rib was to be installed in the upper wing to provide a more stable anchor point for the aileron attachment spar.
3. Improving the joining point between the aileron attachment spar and the wingtip.
4. Strengthening the cut-out for the aileron linkage cable in the aileron attachment spar.
5. Improvement of the joining points between the aileron attachment spar and the rib gussets connected to it.

This aircraft (139/17) was also assigned to Jasta 11 and was the mount of Leutnant von Conta.

6. Improving the connection between the web ribs and the rib gussets.
7. Stiffening the web ribs by glueing additional reinforcing strips of plywood vertically.
8. Removing the lightening holes in the ribs to reduce the collection of moisture.
9. Fabric covering must be sewn to the ribs; no longer would the use of pins be permitted.
10. Measures for strengthening the covering must be undertaken, particularly in the area of the central wing.
11. Wing design must be protected from moisture penetration by the application of a protective coating.
12. The Flugzeugmeisterei is undertaking additional testing of the aerodynamic counterbalance surfaces which may also require modification.

This letter also included the comment that the new wings would have to be delivered without charge. All wings currently in existence were to be taken to ZAK at Adlershof where they would be modified accordingly. Until all Dr.Is were fitted with the modified wings the grounding would remain in effect. The modified wings would have to be certified as good by the Flugzeugmeisterei before any new wings would be accepted for fitting. This writ carried the seal of the commanding general of the Luftstreitkräfte, or Kogenluft, (from KOmmandierender GENeral der LUFTstreitkräfte) and was signed by Hauptmann Mühlig-Hoffmann.

When the United States declared war against the German Reich on 6 April 1917, the Allies could expect an ever-increasing quantity of war materials, which arrived daily at ports along the western coast of France. Because of this, Germany soon found itself facing a massive superpower of materials and soldiers. The air armadas of Germany's enemies also grew by leaps and bounds. For every ten new aircraft Germany built, at the same time a hundred new airplanes would be off-loaded at France's coastal ports.

Since aircraft production in Germany could not compete on the same scale, Army Command searched for another way to effectively counteract the Allied superforce. Accordingly, it was decided to incorporate the most successful fighter squadrons (Jagdstaffeln) within the German Fliegertruppe into specialized units, the Jagdgeschwader (fighter wing).

The first of these units was established on 24 June 1917. This was Jagdgeschwader I.

Quote from the war journal of Jagdgeschwader I:

"According to the telegram from Army Group Kronprinz Rupprecht (Ic 20706), Jagdstaffel 4, 6, 10 and 11 from 4 Army are to be immediately formed into Jagdge-

This is probably 141/17, the randomly selected Dr.I examined at Adlershof after the accidents.

schwader I. The wing is a single-function unit. Its role is to fight for and secure air superiority above critical battle sectors. It remains directly subordinate to KDR 4. Where possible, the individual components of the wing are to be combined at one airfield."

On the following day Manfred Freiherr von Richthofen, who until then had commanded Jasta 11, was appointed commander of this, the first fighter wing in German military history.

In order to have at least three fighter wings available for the major offensives being planned for the spring of 1918, an order from 2 February 1918 called for the immediate formation of Jagdgeschwader II and III.

Jagdgeschwader II was originally to have been formed from Jagdstaffeln 13, 14, 15 and 19 and subordinated to VII Army. The commander of Jasta 12, Hauptmann Adolf Ritter von Tutschek, was called to be its first commanding officer. Understandably, von Tutschek wanted to bring his old staff with him to his new posting as Geschwader commander.

Triplane 144/17 after its capture by British forces. This aircraft was flown by Leutnant von Stapenhorst, whose personal markings in the form of a checkerboard can be seen on the fuselage decking.

For this reason, one of the squadrons was taken from the wing and replaced by Jagdstaffel 12. This was Jasta 14. This unit had received a handful of the new Fokker Dr.Is prior to the official establishment of Jagdgeschwader II. Following its exclusion from this wing, Jasta 14's pilots retained their mounts and, in doing so, this became the only squadron which did not belong to one of the three fighter wings yet continued to operate the triplane from its introduction.

Jagdgeschwader III was formed from Jagdstaffeln 2, 26, 27 and 36. Just like JG I, it too fell under the direct supervision of IV Army. Up until this point in time, Jagdstaffel 26 had been led by Oberleutnant Bruno Lörzer. When he was appointed as the new wing's commander, his brother Fritz Lörzer was made in charge of Jasta 26.

The Triplane's Service with the Jagdgeschwader
Jagdgeschwader I
Initial deliveries of the new aircraft occurred on 12 and 13 December 1917 following the Dr.I's recertification for front-line operations which had taken place on 28 November 1917. These aircraft were assigned to Jasta 11 and were delivered from Schwerin to the IV Army's airpark. They reached Jasta 11 probably sometime in early January 1918.

By mid-February 1918 Jastas 6 and 11 still had not been fully rearmed with the

This photo of 144/17 was taken in the Agricultural Hall at Islington, England. After the aircraft's capture, a detailed study of the design was made and the aircraft became the subject of an official report. Seen in the background are two Pfalz D.IIIs which had also been captured.

new triplanes, while Jastas 4 and 10 had received no triplanes at all up to that point. At this time Jagdgeschwader I was operating no less than three different combat types: the Albatros D.Va, the Pfalz D.IIIa and the Fokker Dr.I.

Jagdgeschwader I was subordinated to II Army on 20 March in anticipation of

the large-scale offensive that same month in order to support its ground activities from the air as well. The so-called March Offensive began on 21 March and lasted until 28 March. Following this particularly successful battle

146/17 shortly after its arrival at Jasta 11 in Avesnes le Sec in December of 1917.

This is also a shot of 145/17. The picture was probably taken in April 1918, since the iron crosses had already been changed to the new Balkan crosses following an order in March 1918. The aircraft now also carries the red markings of Jasta 11, as well as the pilot's personal emblem in the form of a fuselage band.

Leutnant Keseling was forced down behind enemy lines in this triplane, where both he and the aircraft were captured by the British. 147/17 belonged to Jasta 11. A particularly noteworthy detail is the two rectangular openings in the fuselage fabric covering immediately below the upper fuselage longerons, just in front of the machine gun ammunition feed openings. These holes gave access to the anchor points for the center wing.

Jagdgeschwader I was assigned to VI Army in April of 1918 for the upcoming Kemel Offensive. On 20 April 1918 Jasta 4 received a supply of triplanes. Jasta 10 continued to operate without the type, however. The commander of Jagdgeschwader I, Manfred von Richthofen, did not push for further deliveries of the triplane for his Geschwader since he anticipated receiving the newer and better Fokker D.VII at any time. On 21 April 1918 the German nation suffered a tragic loss through the death of one of its greatest heroes, unbeaten in combat 80 times. Freiherr von Richthofen fell in combat while flying his red-painted Fokker Dr.I triplane, 425/17, reputedly killed by the guns of the Canadian volunteer pilot A. Roy Brown.

The Rittmeister von Richthofen left his Geschwader a will which affected its future operation:

"If I do not return, Oberleutnant Reinhardt (Jasta 6) is to assume command of the Geschwader."

And with that, the matter was over.

On 25 April the second large offensive began, its goal to reach the Kemel. By 1 May it was over, and with its conclusion JG I ceased to operate under the command of II Army. The third and last great spring offensive was to begin at the end of May in XVIII Army's sector. Its goal was the river Marne. For this purpose JG I was now assigned to VII Army.

Even before the beginning of the Marne Offensive on 27 May 1918 the first Fokker D.VIIs had begun reaching Jagdstaffeln 6 and 11. At first, these replaced the old Albatros D.Va and Pfalz D.IIIa types, leaving the squadrons to fly a mixture of Fokker Dr. Is and Fokker D.VIIs.

By the start of the Battle of the Marne, however, Jastas 6 and 11 had completely re-equipped with the D.VII.

By mid-July of 1918 Jasta 4 had also fully converted to the Fokker D.VII. The Fokker Dr.Is which had been replaced made their way to the airpark of VII Army, where they were reassigned for continued operations with less important fighter squadrons.

Jagdgeschwader II

This fighter wing was equipped with triplanes even before it was actually established. On 24 December 1917 Jasta 14 received its first deliveries. By 24 January 1918 a total of about 26 Fokker Dr.Is had been delivered to the Jagdstaffeln of Jagdgeschwader II.

On 13 March 1918 the commander of Jagdgeschwader II, Hauptmann Adolf Ritter von Tutschek, fell in combat while flying his triplane 404/17. Hauptmann Rudolf Berthold was appointed as his successor. Hauptmann Berthold was particularly well suited to the post of wing commander since he fulfilled all criteria. He had the rank of Hauptmann(captain), was an experienced and successful front-line pilot and had previously commanded a fighter squadron.

As in Hauptmann Tutschek's case, Hauptmann Berthold wanted to bring his old squadron, Jasta 18, with him to Jagdgeschwader II. However, this time it wasn't as easy to accomplish as Tutschek had found it.

154/17 also belonged to Jasta 11.

One of the most well-known Fokker triplanes was Richthofen's 152/17. This aircraft belonged to Jasta 11 as well.

Tutschek had had the advantage in that Jagdgeschwader II had not yet been officially formed at the time he wanted to bring Jasta 12 along, and it was therefore still relatively easy to swap one squadron for another. Now, however, JG II had been operationally active for over a month and it would no longer be a simple matter. Nevertheless, Berthold got his way with the Army. He achieved this by exchanging all the personnel of Jasta 15 with those from Jasta 18. In doing this, Jasta 15 remained part of Jagdgeschwader II, but now had all of Berthold's people from Jasta 18.

This also had its effects upon the triplanes of the former Jasta 15. Jagdgeschwader II was also subordinated to XVIII Army for the March Offensive. Since the beginning of this offensive was iminent, Berthold also decided to maintain the operational readiness of all the Jagdstaffeln within his Geschwader. He opted not to retrain the new pilots of Jasta 15 on rotary-engined aircraft types and instead determined that they would initially retain their old Albatros D.Va and Pfalz D.IIIa fighters. The triplanes, which had belonged to Jasta 15 up to that time, remained with their Jasta 15 pilot owners when they were transferred over to Jasta 18.

Around the middle of July Jagdgeschwader II began to re-equip with the new and better Fokker D.VII biplanes. In the same month Jasta 19 had, for the most part, converted over to the new fighters and now flew a mixture of Fokker triplanes and Fokker D.VIIs.

Also around mid-June fuel rationing began in an attempt to slow down the immense consumption of this commodity. Each Jagdgeschwader now had just 14,000 liters of fuel and 4,000 liters of oil available to it per month, and from now on would have to carry out operations with this limited supply. It was for this reason that XVIII Army took a drastic step which would severely impact the future of Fokker's triplane. Since inline engines had a significantly better fuel consumption to performance ratio than rotary engines, the decision was rapidly reached to remove the triplanes and their 110 hp rotary engines from service with JG 2. All triplanes were accordingly handed over to the XVIII Army's airpark and from there reassigned to units with less critical roles. The consequence of this was that approximately 40% of all pilots from JG 2 were without operational aircraft. This situation soon changed, however, when deliveries of the Fokker D.VII continued.

Jagdgeschwader III
As with Jagdgeschwader II, some of the individual Jagdstaffeln which would eventually make up Jagdgeschwader III had already re-equipped with the triplanes before they officially became resubordinated to Jagdgeschwader III. An example of this was Jasta 2 (Boelcke) and Jasta 36.

Jagdgeschwader III was also subordinated to the leadership of XVIII Army for the March Offensive. Additional Fokker Dr.Is began arriving following the conclusion of this large-scale offensive. By early April Jasta 26 was operating only the triplanes and Jasta 27 had nearly completed its conversion, flying a mixture of Fokker triplanes and Albatros D.Va types.

With this mixture Jagdgeschwader III took part in the Kemel Offensive from 25 April 1918 to 1 May 1918. Within the next three weeks the first Fokker D.VIIs were already being sent to Jagdgeschwader III. These were first issued to Jasta 27, which now was operationally flying three different types: Albatros D.Va, Fokker Dr.I and Fokker D.VII.

At about the same time as the first Fokker D.VIIs were arriving at Jasta 27, Jasta 2 and 29 were also re-equipped with the new type.

On 24 May 1918 Jagdgeschwader III became resubordinated to VII Army in anticipation of the Battle of the Marne.

163/17 was also in the inventory of Jasta 11.

This triplane, 167/17, bears the squadron markings of Jasta 19. The personal markings consist of a yellow "3" on the fuselage sides and decking. The white field backgrounds on the upper wing have been overpainted in olive-green.

Deliveries of the new Fokker D.VII continued apace, and by the middle of June 1918 nearly all of the triplanes belonging to Jasta 2, 26 and 27 had been transferred to the army airpark of VII Army. By 19 July 1918 only Jasta 36 was continuing to fly the Fokker Dr.I, since there wasn't enough Fokker D.VIIs for a complete conversion to the newer type. A small number of Fokker triplanes remained in service with Jastas 2, 26 and 27 where they found continued employment as reserve aircraft.

Markings and Paint Scheme of the Fokker Dr.I

The factory finish in which the triplane was delivered from the Foker Flugzeugwerke in Schwerin/Mecklenburg consisted of a hand-brushed camouflage paint coat on the upper sides and a light blue-gray color on the aircraft's underside. In addition, the factory painted on the national markings and the iron cross over a white background on the tail, plus added identification markings (i.e. the aircraft serial number, or Werknummer, on all individual components) and the military acceptance numbers on the fuselage sides, to include the weight table on the left fuselage below the cockpit. A so-called data line was also painted on the fuselage side at the factory. This black stripe was an aid for the mechanics during installation of the engine. It was approx. 1 cm wide and showed the engine's exact centerline. The line was 1 meter long and began behind the carburetor air intake on both sides of the fuselage.

The Triplane's Camouflage
General:

The upperside camouflage consisted of a sparingly applied, smeared stripe-like paint application. The camouflage effect was achieved by initially painting diluted light blue-gray in straight lines over the stiffening dope which had been used to treat the fabric covering, so that the original beige color of the linen fabric could still easily be seen. After this coat had dried, the same method was used again to apply an olive-green color. In this manner an impressive three-tone camouflage scheme was achieved.

The direction of the paint strokes did not always have the same angle on each section; the strokes on the fuselage sides were perpendicular to the direction of flight, while those on the upper fuselage decking, the horizontal stabilizer and the elevators were diagonal, applied at roughly a 45 degree angle (from above right to below left).

Leutnant Hans Müller's triplane, 185/17. Both plane and pilot belonged to Jasta 12..

This Fokker, 190/17, was the mount of Leutnant Otto Löffler. Löffler's personal markings consisted of a light yellow fuselage band bordered by two white strips.

The individual components such as fuselage, horizontal stabilizer and elevator were painted separate from each other, and for this reason the above-mentioned approx. 45 degree angle isn't the same overall. Sometimes it was steeper and sometimes shallower. We assume that the reason for this was that the individual parts such as horizontal stabilizer, elevators and ailerons were painted in another workshop than the fuselage and wings. This conclusion was derived from the fact that the stripes on the wings and fuselage were applied at a significantly shallower angle than on those of the aircraft's other components.

On the other hand, it is entirely possible that the painter simply chose the simplest path and kept the stripes as short as possible to finish his work in a rapid manner. By using the lightly-angled strokes the outline of the aircraft blurred into the background quite well, which in turn led to an outstanding camouflage effect. It is not entirely improbable that the varying angles of the strokes wasn't a deliberate part of the overall camouflage.

In detail: The light blue-gray, as well as the olive-green on the upper surfaces was applied using simple continuous paintbrush strokes. Based on photographs and

old pieces of covering which have survived, it is evident that the width of the brushes was around 8 centimeters. Each individual brush stroke continued uninterrupted from the leading to the trailing edge of the part being painted. The brush was not dipped back into the paint after each stroke, but was used until it had nearly run dry. Only then was it immersed in the paint again. Each brush stroke ran right alongside the other. A study of old photos and surviving pieces of fabric covering give no evidence of the brush being rubbed across the fabric or strokes being made in the opposite direction. The Fokker triplanes left the factory with this finish. The undersides were apparently painted in a light blue-gray color.

The same color was used along the edges of the fuselage, horizontal stabilizer and elevator, as well as along the edges of the ailerons, and was a border strip 20 mm wide. The separation line between upper and lower camouflage ran across the middle of the wing leading and trailing edges. The border along the leading edge in particular was often not very well defined. Landing gear struts and cabane braces were painted olive-green at the factory, as were the engine housing and all aluminum covers on the upper side. On the other hand, the wing braces carried the same color as the underside (light blue-gray).

In addition, the wing interplane struts bore black markings which consisted of the Werknummer and the suffix "UL" or "OL" (unten links = below left, oben links = above left), or "UR" or "OR" (unten rechts = below right, oben rechts = above right).

The national markings were applied on the Fokker Dr.I in the standard eight places as well. These were two on the upper side of the top wings, two on the underside of

Leutnant Hermann Vallendor of Jasta 2 flew Fokker Dr.I 195/17. His personal marking on the triplane was a large "V" on the fuselage sides, the decking and the upper part of the top wing.

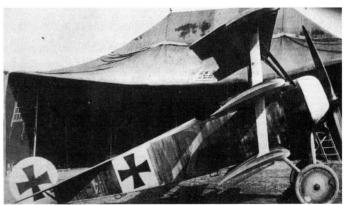

Leutnant Walter Götsch's 202/17 belonged to Jasta 19. The markings on the fuselage side were painted over in olive-green to make room for his personal markings, a yellow "2". From this photograph it can clearly be seen how his personal marking was also subsequently applied in yellow beneath the horizontal stabilizer.

the lower wings, one each on either side of the fuselage and one each on either side of the rudder. Virtually all national markings had a white background, although on a few aircraft the background behind the national markings was left in the beige fabric color. The Maltese cross was used as the national markings right up through the end of the Dr.I's last production run. The Fokker factory painters never deviated from this practice, despite the fact that there were already new field regulations governing the shape and size of the national markings.

The Triplane's Markings

Each military aircraft accepted by the Army was to have clearly been marked and these markings were to have followed specific directives as to their placement on the fuselage sides. In order to avoid confusion, each aircraft manufacturer was given an official abbreviation for his company's products. Examples of these are: "Alb." for Albatros, "Halb." for Halberstadt, "Ru." for Rumpler, "SSW." for Siemens-Schuckert, and so forth. The official abbreviation for Fokker was "Fok.". This abbreviation was a primary component of the specific aircraft identification marking, which also included the aircraft class (e.g. D. for Doppeldecker - biplane, Dr. for Dreidecker - triplane), and the order number under which the Army had ordered it. Finally, this was followed by the year in which the aircraft was ordered. An example of how to read the markings is as follows:

Fok. Dr I 160/17 is a Fokker triplane, type 1, order number 160, ordered in the year 1917.

The position of these markings was also precisely spelled out by regulation and was to be placed low on the fuselage side between the fuselage national markings and the cockpit.

In addition, for identification purposes the aircraft also carried these markings:

1. Components were marked with the serial number of the aircraft. These were usually painted on in black and were located on:
- both fuselage sides immediately ahead of the carburettor intake
- both sides of the rudder between the hinges, above which "Fokker Dr.I" was written
- the damping counterbalance, in the middle over the rudder hinge
- the elevator immediately behind the horns
- the ailerons between the horn and center hinge
- the cabanes, generally just below the wing anchor point on the outside of the forward brace
- the wing braces with the suffixes mentioned earlier in the text
- left underside of the wing with the suffix "Dr.I"
- upper wing between center and cabane

202/17's engine cowling was virtually destroyed by a loose pushrod.

The personal markings on Leutnant Richart Plange's 203/17 included an additional iron cross just ahead of the national markings on the fuselage.

210/17 at Jasta 36. This photo was taken in Kuerne.

This was the method used to transport triplanes from the Fokker works to the front lines.

Possibly 212/17 in the markings of Jasta 2. Leutnant Gallwitz and his mechanics pose in front.

- left center and lower wing next to the fuselage
- forward edge of wing spar with upper edge of script toward wing leading edge
- upper wing topside painted onto wing spar. This was where the celluloid window was cut into the center of the wing. To the left of the window "Dr.I" was painted onto the fabric.

2. Marking of an aircraft's wings with the wing numbers. The wings of a completed aircraft were built together as a single unit and therefore all wings of a particular airframe were assigned wing numbers, which were located on:
- the lower side of the upper wing to the right opposite the Werknummer. In order to better differentiate this number from the Werknummer, the former was accompanied by "Fl.Nr."
- the upper side of the top wing immediately behind the window frame, painted onto the fabric
- the underside of the center and lower wings along the leading edge of the wing spar to the right next to the fuselage

3. Markings on the engine housing were made by using a stamped plate. On the right side of the engine cowling a stamped aluminum identification plate was attached using four rivets. The following text appeared on the plate: "Fokker Flugzeugwerke m.b.H. Schwerin-Mecklenburg". In addition, there were three blank fields on the plate, one for the serial number, one for the aircraft classification abbreviation and the third for the date of completion. Based on these markings, aircraft wrecks could be identified with accuracy since it was highly unlikely that each and every marked component would be destroyed when the plane was shot down.

Additional Front-Line Paint Schemes
Only a small proportion of Fokker's triplanes actually wore the paint scheme which was applied at the factory throughout the length of their operational lives. Instead, many of the aircraft were repainted several times over while serving at the front. Colorful markings of the individual Geschwader and Staffel units were applied on specified airframe parts and personal insignia also found widespread use.

Markings of the Individual Units with the Jagdgeschwadern
Well before 1917 efforts were being made to introduce specific markings for the different Jagdstaffeln within a particular Geschwader. This was, at least in part, so that a pilot flying in formation with other squadrons could quickly identify others of his own squadron. After the establishment of the three Jagdgeschwader units, this practice made even more sense, since now several Jagdstaffeln would be involved in combat at the same time.

One triplane which sported the most impressive personal markings: Leutnant Kempf's 213/17 in full splendor.

When the triplane entered combat operations, most of the Staffel markings were taken over from the earlier Albatros aircraft.

Of the three Jagdgeschwader units in service, only Jagdgeschwader II had its own individual markings to indicate membership in the wing—the white engine cowling.

The markings of those Jagdstaffeln and their triplanes serving with the Jagdgeschwader were as follows:

A) Jagdgeschwader I
Jasta 4
Engine cowling, edges of braces and interplane struts white.
Jasta 6
Engine cowling black, upper and lower sides of horizontal stabilizer and elevators in black and white stripes.
Jasta 10
Engine cowling and edges of braces cream yellow.
Jasta 11
Engine cowling and interplane struts red. A few aircraft also had the edges of braces and cabane struts painted in red.
B) Jagdgeschwader II
Jasta 12
Engine cowling white, fuselage behind the national insignia black.
Jasta 13
Engine cowling white, fuselage behind the national insignia white.

Jasta 15
Engine cowling white, rudder brown with nationality markings on white field.
Jasta 19
Engine cowling white, horizontal stabilizer and elevator were painted in alternating black and yellow stripes, angled to conform to the fuselage outline. The free field in the middle of the horizontal stabilizer was left in the original factory colors.
C) Jagdgeschwader III
Jasta 2 "Boelcke"
Engine cowling black with white front. Rear right fuselage white, rear left fuselage black. The Staffelführer's (squadron leader) aircraft had the pattern reversed. A few aircraft carried a white edge around the rudder.
Jasta 26
Engine cowling and interplane struts black, fuselage from the cockpit back in black and white stripes perpendicular to the direction of flight.
Jasta 27:
Engine cowling, interplane struts, cabane, horizontal stabilizer and elevators on upper and lower sides all yellow.
Jasta 36
Engine cowling blue.
Personal markings of the pilots:
In order to be easily recognized by other pilots in the heat of battle, a few pilots ap-

plied additional markings to their mounts. A few chose meaningful symbols as their personal insignia, while others opted for different-colored fuselage bands or complete paint jobs for their aircraft. If several colors were used, they often were in reference to a pilot's previous units he'd served with. Most of them came from cavalry or grenadier regiments and had volunteered for service with the Fliegertruppe.

One of the best examples of this is probably the fighter pilot hero, Manfred Freiherr von Richthofen, who did not just paint his aircraft in red. He, too, selected the red color in honor of his former regiment, the Ulan Regiment Number 1 of Emperor Alexander III. This color, combined with his enormous success in combat, earned him the nickname of "The Red Baron". It should be mentioned here that this pseudonym was first used in the '30s, and even then not in direct connection with the person of Manfred von Richthofen. He himself titled his book, which appeared while the war was still going on, "Der rote Kampfflieger", or "The Red Fighter Pilot". The term "baron" is used here erroneously, since this title actually never even existed within the hierarchy of the Prussian nobility.

Leutnant Wilhelm Papenmeier is seated here on the cockpit lip of his 214/17. The aircraft belonged to Jasta 2 and is painted in his personal markings consisting of a simple black-white-red band around the fuselage.

This aircraft belonged to Jasta 12, but in this photo wears the squadron markings of the wing commander. Adolf Ritter von Tutschek flew his first front-line combat mission in this aircraft, Fokker triplane number 216/17.

Aircraft of Jasta 12 in front of their tent hangars in Toulis. The airplane in the foreground is Tutschek's 404/17.

But back to the matters at hand. Numerous wonderful and colorfully painted schemes were used on Fokker's triplanes. It's a shame that nowadays nearly all replicas and models are produced in a rather ugly all-red paint scheme.

The Fokker Dr.I in Service Outside the Jagdgeschwader Units

There were only three Jagdstaffeln which were not included in the three Jagdgeschwader and nevertheless flew the triplane in any great numbers. These were Jastas 5, 14 and 34b. Of course the triplane flew with other squadrons, especially near the end of the war, but these were in much smaller numbers and only in isolated cases.

Jagdstaffeln 5 and 34b started to receive the former machines of Jagdgeschwader I in May of 1918 once the latter unit began converting over to the Fokker D.VII.

Jasta 5:

From 17 April 1918 on Jasta 5 was stationed in Cappy, where it was assigned to Jagdgruppe 2 of VII Army. Because of its location, this squadron was in a position to assume control of Jagdgeschwader I's old aircraft. In May of 1918 its old Albatros D.V. airplanes were replaced by the triplanes from Jastas 6 and 11. By mid-May the Fokker Dr.I was operating with success, but was eventually replaced by the Fokker D.VII. From 4 August 1918 onward, the

squadron apparently no longer had any Fokker triplanes left in its inventory, flying the D.VII exclusively.

Jasta 14:

Jasta 14 was the only Jagdstaffel to fly the triplane at the same time this type was still being operated by the Jagdgeschwader units. In January 1918, which was the time during which the squadron was to have been organized under Jagdgeschwader II, the unit received a large number of new triplanes. Shortly afterwards, Jasta 14 was replaced within Jagdgeschwader II by Tutschek's Jasta 12. Jasta 14's squadron leader, Oberleutnant Werner, was able to get his way, however, and the Jagdstaffel was able to retain its triplanes.

Pilots from Jasta 36 pose in front of Fokker 220/17.

This was the ignoble position in which Leutnant Monington of Jasta 15 parked Fokker 401/17.

The Fokker Dr.Is of this squadron were assigned to VII Army from January 1918 until 19 March, when they transferred to XVIII Army in preparation for the March Offensive. When the offensive ground to a halt at the end of that month, Jasta 14 was billeted with Jasta 30 at the airfield in Phalempin in anticipation of the upcoming Kemel Offensive, when it came under VI Army's control. The aircraft continued to operate from Phalempin until 18 August, and up until the sixth of August the triplanes of Jasta 14 flew with the aircraft and pilots of Jagdgruppe VII. This fighter group consisted of Jagdstaffeln 29, 30, 43 and 52.

The last known operation carried out by Jasta 14's triplanes took place on 19 August 1918, when a British report made mention of five of Jasta 14's triplanes becoming involved in an aerial battle. From 30 September onward Jagdstaffel 14 was subordinated to IV Army, belonging to Jagdgruppe III (Jastas 14, 16b, 29, 56) under the command of Oberleutnant Auffahrt. By this time it was no longer using triplanes in an operational role since it was mainly equipped with the new Fokker D.VIIs.

Jasta 34b:

Like Jasta 5, Jasta 34b also received the triplanes discarded by Jagdgeschwader I in May of 1918. The squadron was never brought up to full strength with the Dr.I, however, since by 15 June 1918 the first Fokker D.VIIs began arriving and replacing the triplanes.

In June of 1918 Jasta 34b hardly had any of these small planes left in service. Isolated examples of the triplane were used by various pilots as reserve aircraft or for high-speed, low-altitude intercepts and shooting down observation balloons.

Serious problems began cropping up once again with the engine during the summer of 1918, which rapidly became overheated in the high summer daytime temperatures. Another problem which led to airplanes being removed from operational status was the low quality of the replacement oil used. The last Fokker Dr.I triplane flew in November of 1918, just shortly before the war ended.

Why There Were 320 Fokker Dr.Is

This question is often posed when discussing the little triplane. However, only a few publications have really provided a concrete answer to this.

The matter can be viewed in two ways. First, we can ask why 320 of the Fokker Dr.I aircraft were built at all. Second, we can wonder how the precise number of 320 was determined.

The first question can be answered in the following manner. Two important factors were responsible for the Fokker D.I only being built in a quantity of 320. The first was that the Idflieg anticipated the series production of the Pfalz Dr.I as soon as the SH.III engine had worked out its teething troubles, and second, Fokker himself disqualified the Dr.I because of the problems with the wings. But now we come to the question: Why?

When the Albatros D.V. was introduced into the individual Jagdstaffeln, German pilots could break off combat with a British triplane any time they wanted. This was possible due to the higher speed of the Albatros D.V. in comparison with that of the earlier D.III. This fact was overlooked by the officers of the Kogenluft.

This photo of 404/17 was taken early in the aircraft's operational life. The rear fuselage has not yet been fully painted in black.

Leutnant Götsch and his second triplane, 419/17.

When the Fokker Dr.I appeared at the front in August of 1917, the surprise effect was no longer there. Nor was it able to claim the air superiority that the German pilots were demanding at the front. The British pilots also quickly realized that they could quite easily survive an engagement with the Dr.I by simply breaking off combat, just like the German fighter pilots had earlier been able to escape from the Sopwith triplane. The triplane concept was only successful if the enemy joined combat. And this meant that from a technical standpoint the triplane concept was already obsolete. Climb rate and maneuverability were of no value when the enemy was faster and could flee at any time. This deficiency, coupled with the delay of large quantity deliveries caused by the triplane's grounding due to its wing problems, did not bode well for Fokker's little airplane. Even if the triplane met the requirements of the Army, its technical performance capabilities lagged behind those of the British biplanes. And this was the reason for its short service life and the fact that only 320 examples were built.

In order to answer the question why precisely 320 of the aircraft were built, all we need to do is employ a bit of mathematics. Each Jagdgeschwader (fighter wing) consisted of four Jagdstaffeln, or fighter squadrons. The Jagdstaffeln themselves had a nominal strength of 21 operational fighters each. Ergo, a complete Jagdgeschwader had 4 x 21 aircraft in its inventory, making a total of 84 machines. In addition to these 84 aircraft were reserve planes as well as replacement aircraft for replacing those fighters written off in combat or accidents. This gives us approximately 100 aircraft per Jagdgeschwader. Since only three complete Jagdgeschwader were established before the war ended, the initial inventory of these three fighter wings would logically have been a total of 300 aircraft. And since the Jagdgeschwader were always fitted out with the latest and best fighters as a priority, the Idflieg placed an order for 300 of the Fokker Dr.I machines. In addition to this, there was the initial order for 20 aircraft to be used for front line evaluation, which brings us to the total number of 320.

Without a doubt, Fokker could have sold more triplanes if the matter of the wing breakage hadn't drawn a line through its accounts. For if that hadn't happened, the Idflieg would not only have equipped the Jagdgeschwader units with the nimble little fighter, but would have also ensured that the remaining fighter units of the Fliegertruppe also operated the type.

Technology of the Fokker Dr.I

The Fokker Team Schorndorf is quite familiar with the construction of the triplane, since during the last ten years we've thoroughly researched and reconstructed its design. Based on this, we've built an exact copy of a triplane based on our reseach results to provide historians with a basis for discussion. The copy is on display in the Technik Museum in Speyer, a branch of the Auto & Technik Museum.

This triplane was the mount of Leutnant Heinrich Bongartz, the squadron leader of Jasta 36. One of this squadron's distinguishing marks was the additional iron cross painted on the upper side of the horizontal stabilizer.

A small picture portfolio containing many detailed photos of the copy's construction and design is available directly from the authors (see address at end of text).

At this point we will direct our attention to the technical design of the aircraft itself. In doing this, we will proceed in a systematic way and break down the triplane into its individual assembly steps. These stages are the same ones which were drawn up by the Deutscher Elektronik Dienst during the preparation of the blueprints.

Breakdown is as follows:
Assembly group 100 - powerplant
Assembly group 200 - fuselage
Assembly group 300 - landing gear
Assembly group 400 - horizontal controls
Assembly group 500 - vertical controls
Assembly group 600 - lifting surfaces
Assembly group 700 - equipment and instrumentation
Assembly group 800 - armament
Assembly group 900 - complete assembly

Construction notes
Assembly group 100
As delivered from the factory the triplane was powered by a 110 hp Oberursel Ur II nine-cylinder radial engine. Other engines were also used, however, and included the 100 hp Oberursel Ur I or the 110 hp Le Rhône. During front-line operations it was common practice for pilots to remove the Clerget engines from captured British aircraft and install them on their triplanes, since these powerplants were generally more reliable.

The Oberursel Ur II was, as mentioned above, a radial engine design, meaning that the crankshaft remained fixed while the engine crankcase together with its pistons, etc. rotated around it. Engine and propeller therefore had the same rate of revolution, which in this case was a maximum of around 1200 r.p.m.

The propeller had a diameter of 2,620 mm and was made of different types of laminated wood. Assembly group 100 also included the engine cowling, the splash board (firewall), the engine mounts and the fuel tanks.

The engine cowling was made of 2 mm thick aluminum sheeting and consisted of the shroud, the frontplate and the engine cowl assembly ring. Its role was to cover the rotating rotary engine and protect it from foreign object damage as well. The overall shape of the engine cowling underwent several developmental stages, so that the triplane had about three differently shaped cowls during its operational life.

The splash board is more commonly referred to as the firewall. It is located directly behind the engine and its role is to

The same aircraft taken from another angle. This perspective shows the upper wing's crushed leading edge to good advantage. The aircraft probably nosed over after a crash landing. The motor and its support mounts have been removed.

protect the rear part of the fuselage from becoming smeared with spray from engine oil and fuel, as well as to protect the aircraft from the flames of an engine which has caught fire. The engine in mounted on a large main bearing race. In turn, this race is attached to inward directed star shaped steel tubing. Tube lugs are welded on to each point of the "star", the purpose of which are to serve as guides for the anchor bolts. These in turn are used to bolt the engine support mounts to the corners of the frontmost fuselage bulkhead. By simply loosening these four bolts the entire motor with its mounts can be removed from the airframe. This makes maintenance work considerably easier. Running back from each of these four corner points are steel tubes. Together, they form a pyramid-like assembly and at their back end hold the rear smaller ring mount. A crankshaft is attached to this ring mount. This means that the mounted engine is fixed at two points.

The brother of Rittmeister Manfred von Richthofen, Leutnant Lothar von Richthofen, crash-landed his 454/17 on 13 March 1918. The aircraft was a total write-off as a result.

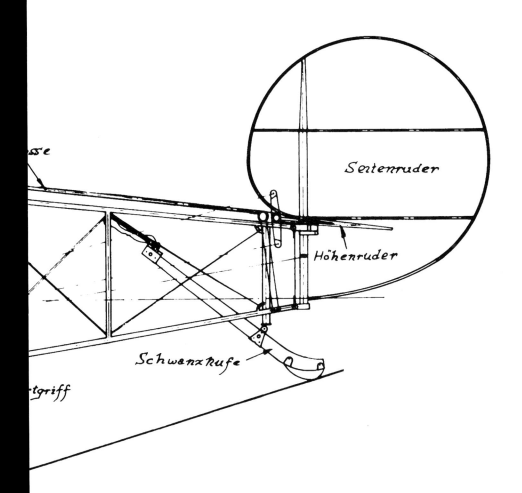

sse

Seitenruder

Höhenruder

Schwanzkufe

tgriff

LÄNGSSCHNITT
Zeichnung: Achim Sven Engels

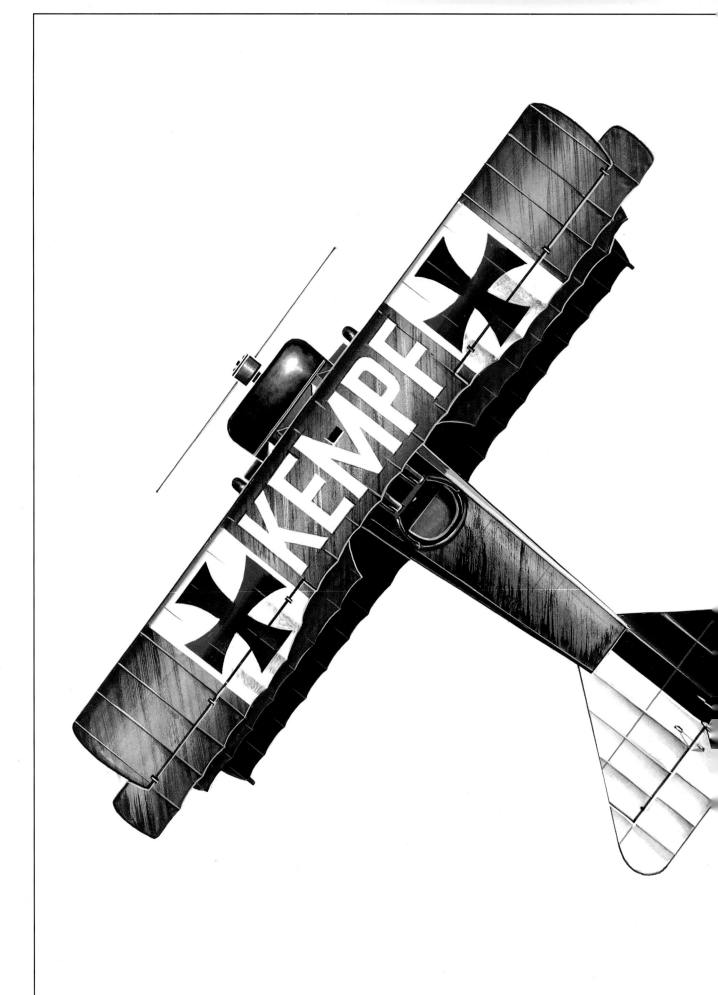

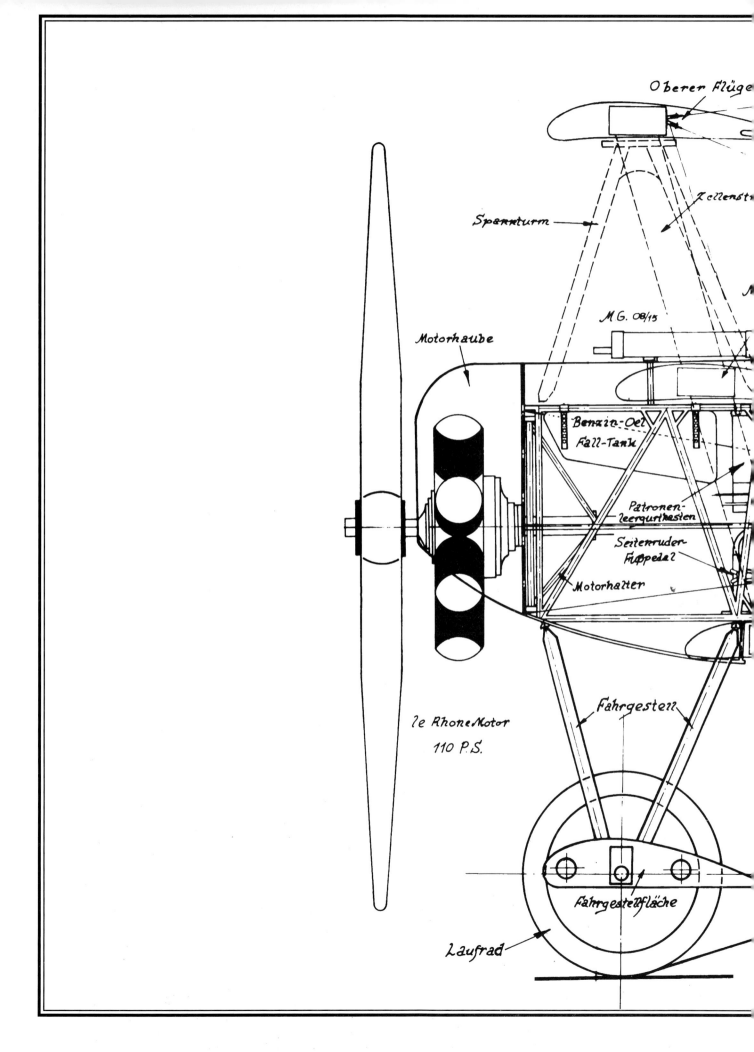

Oberer Flüge

Zellenstr

Spannturm

M.G. 08/15

Motorhaube

Benzin-Oel
Fall-Tank

Patronen-
leergurtkasten

Seitenruder-
Fußpedal

Motorhalter

le Rhone Motor
110 P.S.

Fahrgestell

Fahrgestellfläche

Laufrad

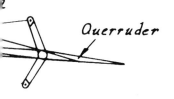

Querruder

ebe

Fokker Dr.I 1917

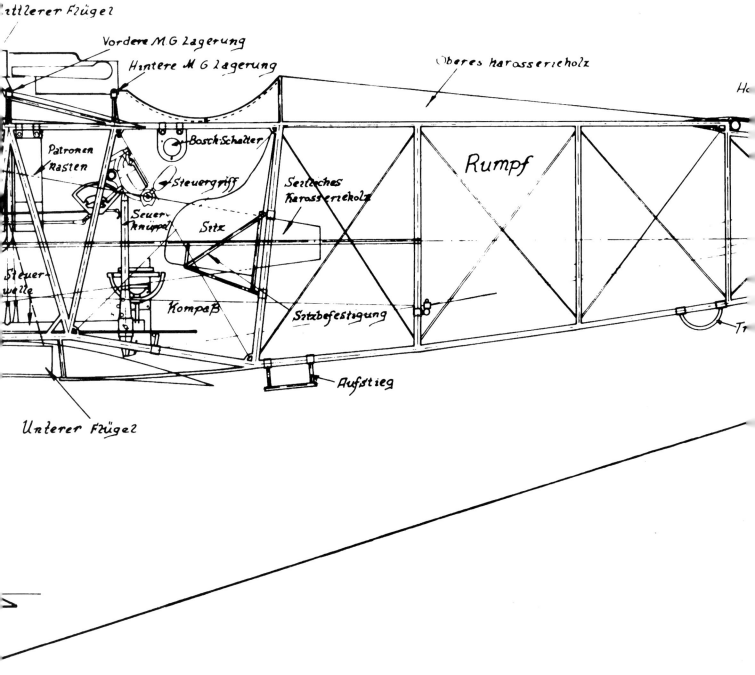

ttlerer Flügel

Vordere M.G. Lagerung

Hintere M.G. Lagerung

Oberes Karosserieholz

Ha

Patronen Kasten

Bosch-Schalter

Steuergriff

Rumpf

Seitliches Karosserieholz

Steuer Knüppel

Sitz

Steuer welle

Kompaß

Sitzbefestigung

Aufstieg

Tr

Unterer Flügel

The fuel tank is capable of holding both fuel and oil for an operating time of 1.5 hours. The tank is built of two chambers separated from each other. The smaller one holds the oil and the second, larger one the fuel. Each of these chambers has both filler and drain openings. Separate lines feed the liquid into the carburettors, where they are mixed together with air. Since the fuel and oil both travel the same route through the crankshaft, an oil must be used which cannot be dissolved by the fuel. In this case castor oil is used.

Assembly group 200

The entire fuselage construction makes use of a meticulously built cross frame design, which is made of welded steel tubing. The cross frame achieves its stability in its rear section through the use of cross-wire bracing. These wires are pulled through small tube eyelets at the junction points of the frame tubing and pulled tight with turnbuckles. This cross bracing ensures a stiff, rugged fuselage frame. In the forward part of the fuselage, this cross bracing is only required on the upper and lower sides, since the latticework frame on the sides is produced in a triangular pattern, providing this area with adequate stability.

The cross section of the forward fuselage is, simply put, a square. This changes into a rectangular cross section as the fuselage continues back. On the underside of the forward square section are found the so called false fuselage longerons. These

Fokker 454/17 seen from the other side after its crash landing on 13 March 1918.

longerons are made of thin steel tubing and bend upward in their center. The underwing spar will later fit into this cutout formed by the bend. The centerwing spar is simply laid on the upper fuselage longeron where it is aerodynamically blended in with aluminum fairings.

The fuselage takes on its rounded cross sectional appearance due to a steel tube curved girder fitted to the foremost fuselage former. This girder serves as the foundation for attaching the side planking, which is made of plywood. These plywood sheets are cut in a triangular pattern and are attached to the steel tube girder by being riveted to gussets made of sheet steel and welded to the girder.

A handful of Jasta 36 pilots pose in front of 465/17. (Left to right) Vizefeldwebel Patzer, Lt. von Häbler, Vizefeldwebel Meyer, Lt. Wedelt, Squadron Leader Lt. Bongartz, Lt. Naujok, Lt. Fuhrmann and Uffz. Hübner.

This machine, Fokker 466/17, belonged to Jasta 14 and wears that unit's squadron markings. Notice the design of the national insignia.

Leutnant Kempf's second aircraft, 493/17. Like before, he decorated this airplane with his personal markings and the slogan "Kennscht mi' noch?" (Remember me?). Unlike 213/17 this triplane already has the Balkan cross painted on.

So that the stretching of the fabric covering doesn't push the plywood inward, the planking has a support located between the tubular frame and the planking itself. This support is fitted along the centerline of the planking, where a wooden board running crosswise through the forward fuselage is pinned and glued to the plywood. Two additional boards are to be found at the planking's edges, the center board jutting out over the planking to the rear. By using a sheet metal gusset, its rearmost end is attached to the vertical steel tube of the first former at a point behind the cockpit.

At the rear end of the fuselage, the vertical stabilizer is attached to the two upper and one lower fuselage longeron with two rudder hinges. The fuselage's stern post is made of wood. The upper side of the rear fuselage is stepped down, and the horizontal stabilizer's frame is placed into the resulting depression, where it is bolted to the fuselage with three screws. Two are located at the forward end of the step, one at the rear.

The tail skid is hung from upper tube junctures of the next-to-last fuselage former by means of elasticated rubber shock

chords. Just ahead of the fuselage stern post is a section of steel tubing welded vertically, with a device located at its lower end to which the skid is attached. This device permits the tail skid to pivot up and down. The overall design enables the skid to effectively absorb shocks which occur during takeoff and landing.

The anchor points for the landing gear and cabane struts, on which sits the upper wing, are also found in the forward section of the fuselage.

This Fokker 525/17 was one of the triplanes used by Manfred von Richthofen. The airplane belonged to Jasta 6.

Fokker 528/17 originally belonged to Jasta 6 and was later assigned to Jasta 5, where its squadron markings were overpainted. Vizefeldwebel Fritz Rumey was the author of this crash landing.

A beautiful side view of Fokker 581/17 with a factory-applied paint scheme.

These anchor points are small lathed parts, also called ball cups. At one end these ball cups have been hollowed out, while the other end has a long shaft pin. This pin fits into holes which have been drilled into the steel tubes of the frame construction, where they're welded solid with the frame. The connecting points at the ends of the landing gear and cabane struts fit snugly into the hollowed out areas of the ball cups, where they're secured by split pins.

A view of the cockpit area of 581/17. With the left cockpit fairing removed, the deviation table and its location can clearly be seen. A sample of this table can be found elsewhere within the pages of this book.

The fuselage provides the basis for all other subassemblies of the Fokker triplane.

Assembly group 300
The landing gear primarily is comprised of the struts, the undercarriage "knees", the crossbrace, the axle airfoil cover, the axle and the wheels.

The undercarriage struts are made from teardrop profile steel tubing. Between the two forward strut legs are the cross bracing wires running diagonally, designed to prevent the landing gear from collapsing. Welded iron attachment points are found on the upper end of the steel tubes, which in turn are attached to the anchor plates on the fuselage. The lower end of the struts are welded to the so-called undercarriage "knees".

These "knees" are box-shaped components made of welded sheet steel. In the center of each knee is an oval vertical opening through which runs the axle. Within this opening, the axle can slip up and down a certain amount. Two steel sleeves are welded to the outside of the knees, and the axle runs through these. Shock to the axle is absorbed by a rubber cord which is wrapped around the two sleeves and the axle. Small frames made of steel sheeting and containing several drilled holes are welded to the inner sides of the knees. The undercarriage crossbrace is attached to these frames using small rivets.

The undercarriage crossbrace is a two-piece aluminum box located between the undercarriage knees, the purpose of which is to keep the landing gear braced horizontally. At the same time, it serves as the attachment point for the axle airfoil cover.

The axle airfoil cover is also a two piece construction which looks like a miniature wing. It is made up of a leading edge, or nose piece, and a rear section. Both the nose piece and the rear section are made from wing ribs which give it its airfoil shape, although unlike the wings, these are covered in plywood instead of fabric.

33　This is a very informative photo showing the wreck of 591/17. The details of the cockpit interior can clearly be made out. Note in particular the shape of the control grip. This one is a later development and is different from the grip used in Richthofen's 425/17 and now on display in the Australian War Memorial in Canberra.

This triplane probably belonged to Jasta 4

Thick aluminum tubes serve as the spars for each of the airfoil's sections, over which slide the ribs. A specially designed wooden frame provides the necessary stability on the open ends and enables both sections to be slid over the undercarriage crossbrace. Unlike the later Fokker D.VII, the axle airfoil cover on the Dr.I could not be removed when the airplane underwent maintenance. In such cases, it was destroyed and replaced by a new one.

The profile of the cover provided the aircraft with an aerodynamic advantage in that it reduced the frontal drag of the undercarriage and offset a good portion of its weight by the airfoil's generation of lift.

The axle of the landing gear is a large steel tube with thick walls. It runs the entire breadth of the undercarriage and is not, as many other publications claim, a two-piece affair which is rigged in such a manner that each half can swing upward around a pivot point; it is instead an all-through unit made of a single steel tube.

The airplane's wheels have a diameter of 700 mm and a width of 80 mm. Each wheel rim has 16 spokes, giving the wheel set a total of 32. The tires are non-treaded. The wheels are attached to the axle through the fitting of a set collar on each side of the wheel. In order to reduce the aerodynamic drag somewhat, the sides of the rims are covered in fabric.

Assembly group 400
Horizontal control surfaces consist of the horizontal stabilizer and the elevator control surfaces. Both pieces are made of steel tubing. The horizontal stabilizer includes a rigid trapezoidal framework made of 35 mm steel tubing. The rear section of this framework is also made of steel tubing. A three-part trailing edge strip made of wood is attached directly to this steel tubing using small aluminum sheet strips.

Hauptmann Wilhelm Reinhardt, Richthofen's successor, poses in front of a triplane from Jasta 6.

The rear ends of the horizontal stabilizer ribs are welded to the steel tubing of the trailing edge. At those points where they cross over the framework, the ribs are welded to short tubular sleeves which are slipped over the tube frame. This feature prevents the frame from becoming weak through the heat generated when welding the ribs.

The ribs on the upper and lower sides converge again just ahead of the frame, where they are welded to the leading edge of the horizontal stabilizer (which is also made of steel tubing). The horizontal stabilizer is braced to the fuselage by a teardrop profile steel tube running from its lower side to the fuselage base.

The elevator has a continuous spar made of 30 mm diameter steel tubing. On the outermost edges of the elevator are the small areas of the aerodynamic counterbalances, located ahead of the elevator spar. These are made by extending a rib forward and continuing through with the elevator's leading edge. The purpose of these counterbalances is to relieve some of the load on the pilot's muscles when operating the elevator. Near the middle of the elevator spar are two elevator control levers made of sheet steel, to which are attached the control cables running from the control stick. They are designed in such a manner that when the control stick is moved forward the elevator moves downward, and when the stick is pulled back it raises the elevator. The control cables are fed through the length of fuselage solely by means of pipe clamps welded at right angles to each other.

Hauptmann Reinhardt with his heavily damaged triplanes after a successful landing behind German lines.

Three elevator hinges act as the attachment points for the elevator to the horizontal stabilizer. These consist of sheet steel bent into a simple "U" shape, which are wrapped around the spar and bolted to the horizontal stabilizer.

Assembly group 500

The construction of the rudder closely mirrors that of the elevator. The rudder also makes use of an aerodynamic counterbalance in front of the rudder spar. Two hinges are used to attach it to the rearmost ends of the upper and lower fuselage longerons. It is operated by a rudder control lever. The frame edge is welded at its bottom to the rudder spar, from where it is bent in an oval pattern to form the shape of the rudder itself. Just ahead of the lower end of the aerodynamic counterbalance region the frame edge splits into two slender steel tubes which then pass around either side of the rudder spar and act as a rib.

The second rib is located in the upper third of the rudder. It is welded to the spar on either side and converges fore and aft, where it joins the frame edge. In the area behind the rudder spar the ribs are reinforced through the use of thin steel tubing arranged in a triangular weave pattern.

Mechanics of Jasta 11 at work on one of their squadron's triplanes.

Assembly group 600

The triplane's wings are made almost exclusively of wood. Only the fittings and ailerons are made of metal. Since the construction of the upper, middle and lower wings is virtually the same, we will only provide a brief overview of each.

The basic foundation for all three wings is the large wing spar. This consists of two box spars joined tightly together. Each box spar consists of two spruce strips, one above and one below. These are joined together from the right and left sides using thin plywood sheeting.

In the same manner the box spars are joined underneath each other. Plywood sheeting on the upper and lower sides holds them together, thus creating a stabile box spar. Their dimensions are 100 mm x 200 mm. The designer layed out the height of the spar in such a manner that it could form an equivalent moment of inertia for the complete unbraced wing.

The ribs of the wing have the same profile and the same design along the entire span for all three wings. They consist of a rib crosspiece made of thin plywood. Rib flanges made of pine are attached to the crosspiece above and below and run the entire length of the ribs. To save weight the rib crosspiece has round cutouts. The rib arches are reinforced in three places with plywood strips glued on their right and left sides. At the point where the spar feeds through the rib crosspiece, a large cutout has been made and reinforced with plywood.

The ribs are simply set vertically on the wing spar and held in place by small triangular wooden fillets. The fillets are laid down against the rib flange from left to right, where they are attached to the spar strips using pins and glue.

At the rear of the ribs is a cutout for the rear auxiliary spar. This is a squared

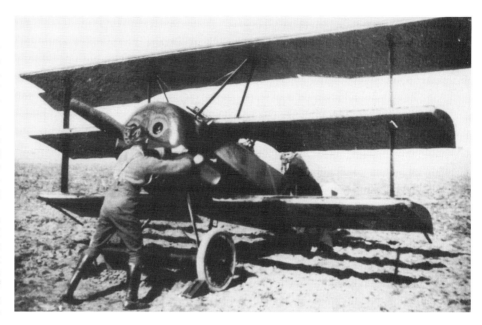

Leutnant Steinhäuser of Jasta 11 decorated his triplane with a face on the engine cowl similar to that which Werner Voss used. A red band and yellow cross adorn the fuselage.

length of wood the purpose of which is to ensure that all ribs remain in their place and not vibrate or shift at their ends. At the rearmost tip of the ribs are small aluminum sheet metal fittings which cover the rear rib tips, where they hold the wing endwire in place. The endwire in turn forms the trailing edge of the wing. The fabric covering is also wrapped around this, and it is the covering which causes the characteristic scalloped pattern of the wing's trailing edge.

The wing's leading edge is covered in thin veneer, which in this area ensures that the fabric covering doesn't fall down in between the ribs. On the upper side the veneer is cut into a triangular pattern, with its uppermost corner being pined and glued to the spar. On the wing's underside the veneer covering ends in a straight line.

A cloth strip weaves through the ribs just behind the wing spar. It laces from the upper side of one rib to the lower side of the next and so on. The fabric strip prevents the ribs from vibrating in flight, which may cause the fabric covering to come loose.

The control cables for moving the upper wing ailerons are fed through holes in the upper wing. The ailerons themselves have an aerodynamic counterbalance on their outer ends.

The new triplane at Jasta 11 on the airfield at Lechell.

The ailerons are attached to a sturdy spar within the wing, called the aileron anchor spar. This spar is fixed to the ribs.

The ailerons are constructed in the same manner as the elevator and rudder. The rudder lever in the cockpit is the only thing used to control them. Like the wings themselves, the trailing edges of the ailerons are formed by a simple stranded wire.

The middle wing has cutouts, located right next to the fuselage, to permit better visibility for the pilot while on the ground. These cutouts are formed by using curved edging made out of laminated wood strips, which run from the inner connecting rib to the next full-size rib. In order to provide the rearmost cabane strut enough room, the rear part of the inner connecting rib is angled sharply outward.

The method of angling a rib segment is also employed on the lower wing in order for the forward part of the lower wing's inner connecting rib to allow the rear landing gear strut to pass through.

The wings are attached to the fuselage by sheet steel fittings bent into the shape of a "U", which have a threaded piece welded at one end. The fittings are screwed down tightly to the wing spar. The threaded end is fed through an eye located in the fuselage frame, where it is also bolted down. The attachment eyes for the upper wing are located in the upper end of the cabane struts. All bolts used in the triplane's construction make use of split pins to prevent them from coming unscrewed by themselves.

Assembly group 700

The triplane's equipment and instrumentation includes all components which the pilot needs in order to operate the aircraft: the seat, the cockpit flooring, the joystick, the camshaft, the rudder pedals, the engine controls and the instruments.

The airplane's seat rests on a frame made of welded steel tubing. This framework is vertically adjustable so the pilot can position himself at a comfortable level. The seat itself is attached to the frame using rivets and bolts and consists of a wooden base to which the aluminum backing is riveted. The seat is upholstered exclusively with a layer of felt padding. Seatbelts are attached to U-shaped rounded bars which rise up through drilled holes in the seat base where they're held fast by nuts. In addition, their anchor points sit behind fuselage formers, where the ends are bent over and joined to each other.

A reflexive gunsight was experimentally installed in Leutnant Graven's triplane. The way the belt harnesses are hanging down toward the machine guns leads one to believe that the aircraft was positioned on its nose especially for this photograph. Quite unusual!

This photo was taken in Toulis and shows the different types of aircraft operating with Jagdgeschwader II at this point in time.

A beautiful shot of a triplane from Jasta 12. Photographs of triplanes in flight are rare.

Leutnant Hans Müller posing in front of an unidentified triplane, possibly belonging to Jasta 15.

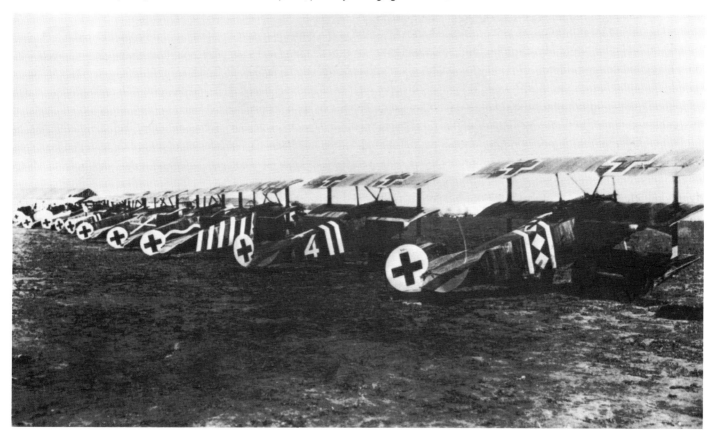

Jasta 19's aircraft in Balatre. The triplane to the far right is Leutnant Arthur Rahn's Fokker 433/17.

Vizefeldwebel Otto Esswein is seen here standing in front of 426/17. Both photos on this page show lineups of Jasta 26's aircraft based in Erchin.

The cockpit flooring provides the pilot with a surface to stand on and is divided into two parts. It is made of 6 mm thick plywood and is bolted to the lower steel tubing of the fuselage formers using small tube clamps. Two curved aluminum plates are located on the topside of the flooring, which are fitted with wooden runners on either side of each plate. These are the so-called footrests and prevent the pilot's heels from slipping when pushing on the rudder pedals. The control stick is made of hollow steel tubing and steel plate. The control column, made of thick steel tubing, is welded to a conical sleeve formed from a piece of sheet steel, which in turn is welded to a larger piece of bent steel tubing running lengthwise along the aircraft's longitudinal axis. This piece has a hole drilled through the middle of the arch through which the anchor bolts for the control stick. Another conical sleeve made of sheet steel can be found on the underside of this curved steel bar. Both sleeves, the upper and the lower one, contain the attachment points for the elevator control cables. The control grip is located at the upper end of the control column. It is made from a specially formed tubular frame and contains the most important control functions. On the right side is a fixed wooden grip. On the left is a second wooden grip, but this one is able to move forward and backward and acts as the throttle lever for the engine.

Between these two wooden grips is the push button for firing the machine guns and the interrupter button for shutting off the engine.

The control shaft runs horizontally along the cockpit floor in the direction of flight. It is free-turning and has at its rear end the anchor bolt for the control column, enabling the control column to be moved freely both right and left. Two rudder levers are positioned at a 100-degree angle to each other and are located in the center of the control shaft. At their ends, these have the attachment points for the aileron control cables.

The rudder pedals are pivotable around the rudder bearing shaft. This is a steel shaft bent backward, which on its top and bottom ends is attached to the middle of the fuselage transverse frame. In turn, the rudder pedal vertical bearing shaft is slid over its lower, straight section. The rudder pedals basically consist of a long steel bar horizontally welded to the vertical bearing shaft. Stirrups made of thin steel tubing are attached to both sides of this rudder bar, which enable the pilot to better keep his feet on the controls. The connection points for the rudder control cables are also found on the pedal bar.

The engine controls are located on the joystick and on the left side of the cockpit, to be operated using the left hand. The carburettor lever and the so-called auto-release lever are attached to a device affixed to one of the vertical fuselage formers on the left side of the cockpit.

Oberleutnant Erich Löwenhardt standing in front of a triplane.

No information is available for this triplane.

A few triplanes were used in flying schools to train pilots who were converting over to rotary engine aircraft. This was the case with this particular machine, painted with a large yellow "74" on the side.

An unidentified triplane is being viewed with interest by German ground troops after making a successful crash landing behind friendly lines.

One of these levers, when twisted, controls the fuel flow valve on the fuel tank. The other is designed to be tilted forward and regulates the flow of the combustion mixture within the carburettor.

The Fokker triplane's instrumentation consists solely of a compass, a fuel gauge and the Bosch ignition switch. Very few examples also had a tachometer, and just as few had an anemometer for measuring airspeed, installed on the strut braces. In many previous publications the triplane has been portrayed as having an instrument panel and a suite of instruments. We would like to emphasize at this point that those rotary-powered aircraft which Fokker built generally had nothing of the sort. On the other hand, the Fokker D.VII had an instrument panel with some instrumentation. The view that the Fokker triplane had a good instrument layout is a popular one, but is not accurate.

Assembly group 800

The triplane's weapons consisted of the standard German aviation armament, the LMG 08/17. This was an air-cooled version of the water-cooled MG 08/15. Its rate of fire was approximately 800 rounds per minute. Two of these machine guns were installed in the upper fuselage, fixed to fire in the direction of flight and, thanks to the control system developed in the Fokker works, were able to fire through the rotating propeller arc.

No information is available for this triplane, either.

Based on the markings, this aircraft may be a Fokker from Jasta 12.

The black and white markings on the fuselage indicate that this triplane belongs to Jasta 26.

This triplane is parked in front of a makeshift hangar made of wood. Such shelters were built to protect the aircraft if a squadron was expected to be based in a particular area for an extended period.

No information has been provided for this airplane. However, the pilot in the cockpit is Leutnant Werner Voss. Interestingly, this particular machine has been fitted with a supplemental headrest made of sheet aluminum with leather upholstry.

This triplane was used in 1938 by Karl Ritter to shoot air-to-air scenes for his film "D III 88".

The Fokker triplane carried 500 rounds of ammunition for each gun. Belts for these were located in aluminum ammunition boxes carried beneath the guns themselves. The rear box held the belts containing the ammunition, while the forward one collected the belts and spent casings. The belt feed was located on the upper part of the rear box.

Assembly group 900
This is satisfactorily covered in the diagrams and drawings included in this book.

Translated from the German by Don Cox

Front cover artwork by Jim Dietz

Printed in China.
ISBN: 0-7643-0400-3

This book was originally published under the title,
Flugzeug Profile: Fokker V.5/Dr.I-Die Geschichte des Fokker-Dreideckers
by FLUGZEUG Publikations GmbH.

We are interested in hearing from authors with book ideas on related topics.

Published by Schiffer Publishing Ltd.
4880 Lower Valley Road
Atglen, PA 19310
Phone: (610) 593-1777
FAX: (610) 593-2002
E-mail: Schifferbk@aol.com
Please write for a free catalog.
This book may be purchased from the publisher.
Please include $3.95 postage.
Try your bookstore first.

Technical Data for the Fokker Dr.I

Powerplant: Oberursel Ur II 9 cylinder 110 horsepower rotary engine. A few triplanes were also fitted with a license-built copy of the Oberursel Ur II built by Rhenania Motorenfabrik in Mannheim.
Le Rhône 9 cylinder 110 horsepower rotary engine.
Göbel Gö II 7 cylinder 100 horsepower rotary engine in those aircraft destined for training units.

Propeller:

Length:	2620 mm
Pitch:	2300 mm
Blade width:	230 mm
Hub:	8 holes
Lamination:	3 layers of walnut, 4 layers of birch

Dimensions:

Upper wingspan:	7190 mm
Middle wingspan:	6225 mm
Lower wingspan:	5725 mm
Wing chord:	1000 mm
Interval between upper and middle wings:	875 mm
Interval between middle and lower wings:	855 mm
Aileron span:	2500 mm
Aileron chord:	300 mm
in region of counterbalance:	500 mm
Horizontal stabilizer span:	2020 mm
Horizontal stabilizer chord at fuselage junction:	1185 mm
Elevator span:	2620 mm
Elovator chord:	425 mm
in region of counterbalance:	778 mm
Fuselage length overall:	5770 mm
Length of fuselage frame:	4490 mm
Height of aircraft:	2950 mm

Surface area:

Upper wing:	7.58 m²
Middle wing:	5.04 m²
Lower wing:	4.86 m²
Axle airfoil cover:	1.20 m²
Horizontal stabilizer w/ elevator:	2.70 m²
Rudder:	0.66 m²
1 Aileron:	0.80 m²

Specific loading:

Wings at full weight:	35.5 kg/m²
Weight to engine performance ratio:	5.2 kg/hp

Angles of incidence:

All three wings:	inner 2.3°, outer 2.5°
Axle cover:	2.0°
Horizontal stabilizer:	4.7°

P49:

Performance:

Takeoff run:	50 - 100 m
Landing run:	50.0 m

Time to climb (1000 m): .... approx. 2.9 min.
Time to climb (2000 m): .... approx. 5.5 min.
Time to climb (3000 m): .... approx. 9.3 min.
Time to climb (4000 m): .... approx. 13.9 min.
Time to climb (5000 m): .... approx. 21.9 min.

Results of speed measurements taken at Adlershof on 9 April 1918

barometric pressure		temperature	air density	rate of climb	actual altitude	horizontal speed
mm Q.-S.	°C	Y	m/sec	m	m/sec	km/h
538	-5	0.93	4.00	2780	43.6	157
505	-7	0.88	3.80	3200	40.9	147
443	-12	0.79	2.20	4180	38.8	139

Weights:

Capacity: .... 203 kg
(pilot 80 kg, guns 22 kg, ammunition 25.6 kg, special equipment 11.4 kg, fuel 64 kg)
Takeoff weight: .... 586 kg

Empy weights for engine:

Engine .... 147.0 kg
Propeller with hub: .... 17.0 kg
Tank system: .... 12.9 kg
Engine accessories: .... 16.2 kg
Total weight: .... 194.1 kg

Empty weights for aircraft:

Fuselage: .... 47.3 kg
Seat, struts, fairings: .... 17.5 kg
Landing gear: .... 34.3 kg
Tailskid: .... 1.0 kg
Controls: .... 3.2 kg
Wings: .... 90.0 kg
Empennage: .... 12.5 kg
Fuselage covering: .... 6.1 kg

Serial numbers:

Acceptance numbers	Factory serial numbers(Werknummern)
100/17	1830
101/17	1697
102/17 - 103/17	1729 - 1730
104/17 - 119/17	1772 - 1787
121/17 - 140/17	1832 - 1851
141/17 - 170/17	1853 - 1882
171/17 - 200/17	1889 - 1918
210/17 - 220/17	1920 - 1939
400/17 - 429/17	1984 - 2113
430/17 - 459/17	2055 - 2084
460/17 - 489/17	2086 - 2115
490/17 - 519/17	2117 - 2146
520/17 - 549/17	2188 - 2217
550/17 - 597/17	2220 - 2267
598/17	unknown
599/17	1919

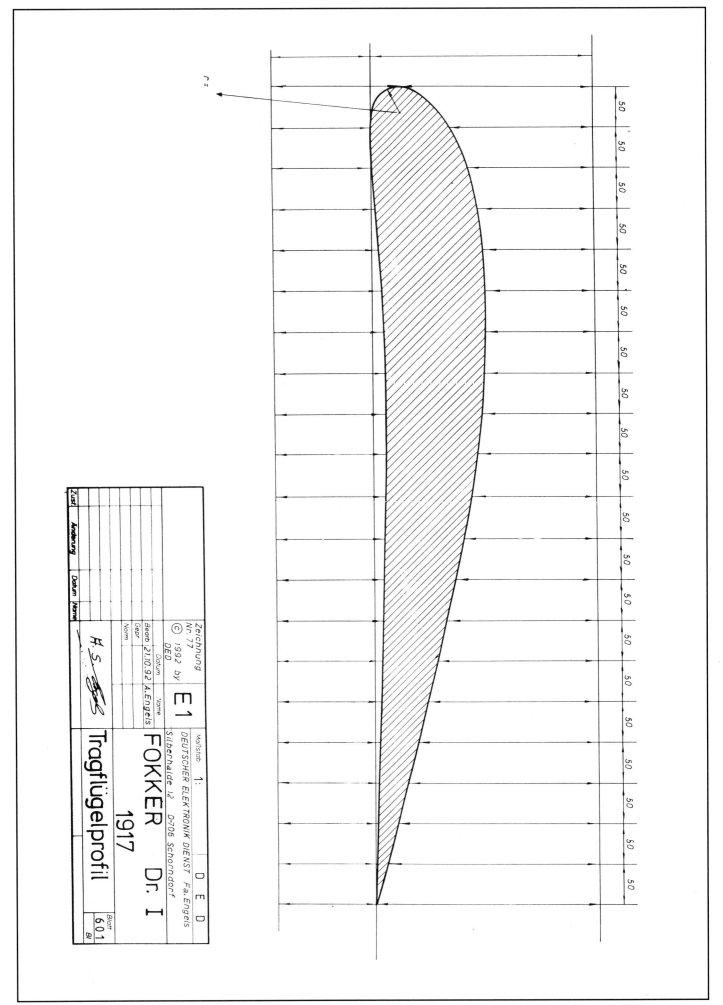

r =

50 50 50 50 50 50 50 50 50 50 50 50 50 50 50

Zeichnung
Nr. 77
© 1992 by
DED

DEUTSCHER ELEKTRONIK DIENST Fa. Engels
Silberhalde 12 D-706 Schorndorf

Maßstab	1:	D E D

	Datum	Name	E 1
Bearb	21.10.92	A.Engels	
Gepr			
Norm			

FOKKER Dr. I
1917
Tragflügelprofil

Zust	Änderung	Datum	Name

Blatt	6 0 1
Bl	

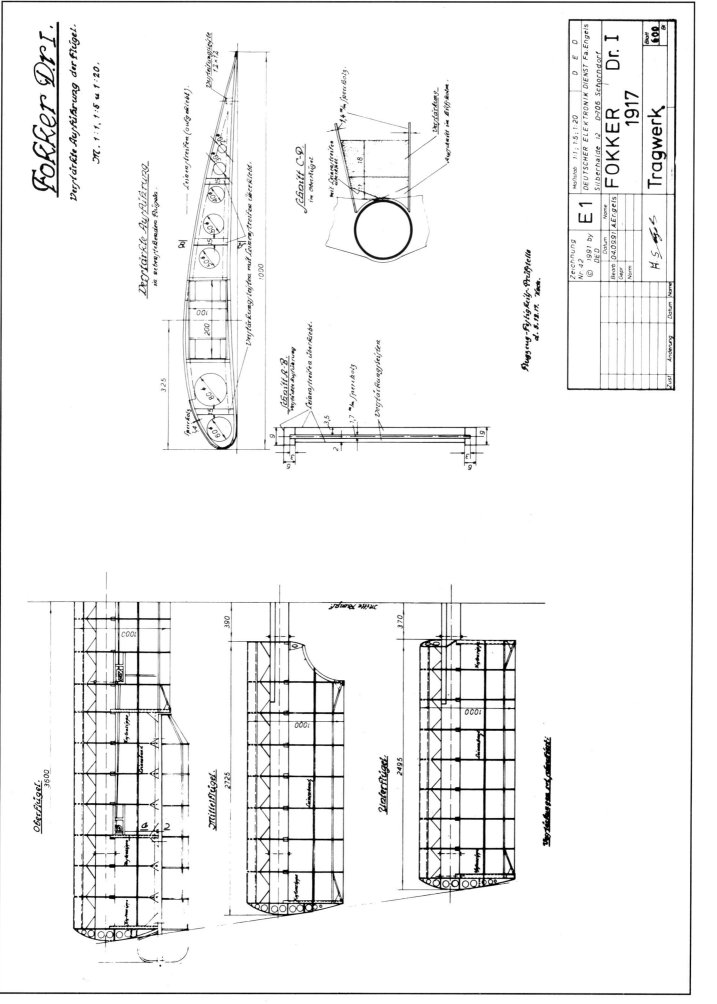

Fokker Dr.I.

Verstärkte Ausführung der Flügel.

M. 1:1, 1:5 u 1:20.

Verstärkte Ausführung
in verbesserten Flügeln.

Schnitt A-B.
Verstärkte Ausführung

Schnitt C-D.
im Oberflügel.

Zeichnung Nr. 42	E 1	D E D
© 1991 by DED		DEUTSCHER ELEKTRONIK DIENST Fa.Engels
		Silberhalde 12 D-706 Schorndorf

	Datum	Name
Bearb	04.09.91	A.Engels
Gepr		
Norm		

FOKKER Dr.I
1917
Tragwerk

Blatt 600

Flugzeug-Festigkeit-Prüfstelle
d. 5.12.17. Kecke.

| Zust | Änderung | Datum | Name |

Oberflügel.

Mittelflügel.

Unterflügel.

Dritte Rumpf.

49

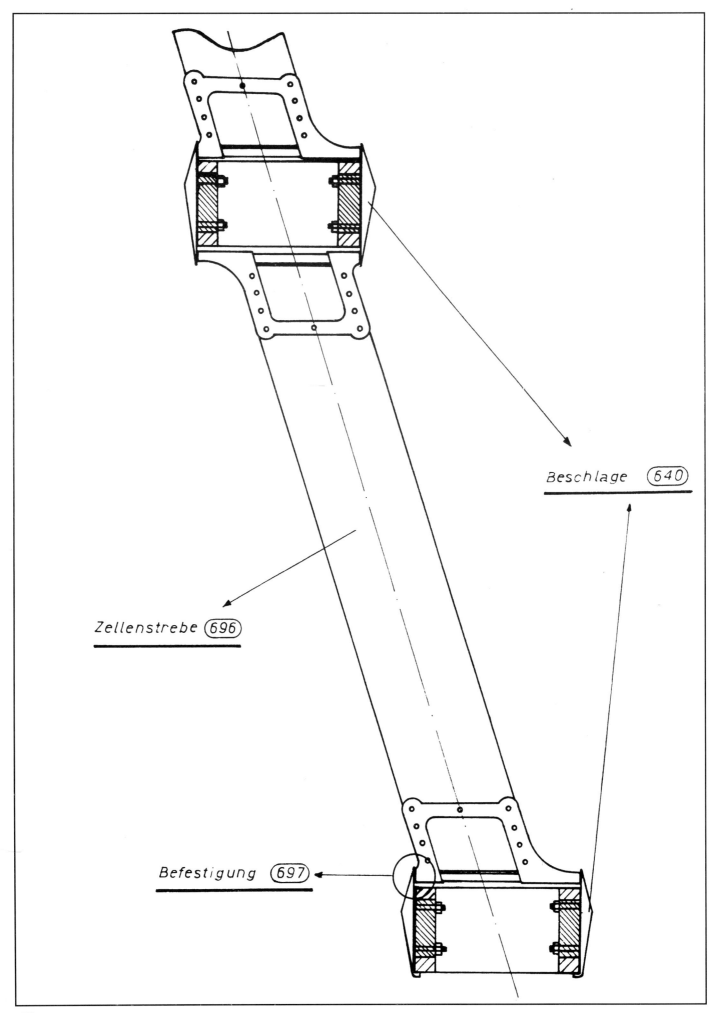

Beschlage (640)

Zellenstrebe (696)

Befestigung (697)